Start a Butterfly Garden

by
Peter Cawdell

with
illustrations by
Stuart Parfett

School Garden Company

Starting a  Butterfly Garden

by Peter Cawdell
ISBN 1-85116-801-X

First Published © 1987 by **School Garden Company,**
P.O. Box 49,
Spalding,
Lincs. PE11 1NZ.
Telephone 0775 - 69518

in association with   **The Butterfly Park,**
Long Sutton,
Lincs. PE12 9LE.
Telephone 0406 -363833

Illustrations by Stuart Parfett

Peter Cawdell is a keen butterfly enthusiast and conservationist, with most of his large garden devoted to encouraging butterflies. He founded the Lincolnshire branch of the *British Butterfly Conservation Society* in 1984 and is currently their branch Chairman.

The Butterfly Park, one of Britain's largest walk-through tropical Butterfly Houses, is owned by Peter and Julie Worth — both keen conservationists. The Park also includes an outdoor wildflower meadow, a butterfly and bee garden, and an aquatic/pond area.

"I only ask to be free. The butterflies are free."
. . . . **Charles Dickens (from** *Bleak House***)**

Contents

One . . . . Introduction . . . . Page 4
Two . . . . Choosing the site . . . . Page 5
Three . . . . . . Butterfly borders . . . . Page 7
Four . . . . Wild areas for butterflies . . . . Page 10
Five . . . . Gardens and butterflies . . . . Page 14
Six . . . . Butterflies — notes on natural
history . . . . Page 22
Seven . . . . . . Some activities and experiments . . . . . Page 25

Appendices

A   Plants to attract butterflies . . . . Page 27
B   Plants for egg-laying and cater-
pillar feeding . . . . Page 29
C   Selected book list . . . . Page 31
D   Useful resources . . . . Page 32
E   Contact addresses . . . . Page 32

Illustrations

Page 4   Garden preparation
Pages 6/7   The two types of butterfly garden
Page 9   Two butterfly borders — plans and plants
Page 11   Coping with nettles
Pages 16/17   Ten common butterflies and their favourite plants
Page 22   Typical life cycle
Page 24   Life cycles of four butterfly species
Page 26   Butterfly breeding experiment

1
Introduction

Butterflies have an almost universal appeal. Their attractive wing colours and graceful flight are associated with warm summer days from early childhood. The idea of deliberately trying to attract these "mobile flowers" into the garden is a relatively new one. An increasing number of gardeners are willing to sacrifice some of the traditional standards of neatness and formality in their gardens in exchange for the beauty and interest these insects will conjure.

As more and more wild places are lost to modern farming practices and to urban development, butterfly habitats diminish daily, and by providing even small areas where butterflies may feed and possibly breed we are helping them to survive for the enjoyment of future generations. These wildlife gardens — perhaps tiny in themselves — but dotted all over the country can add up to a vast nature reserve.

Several butterfly farms with enormous walkthrough greenhouses have sprung up within the last few years. These have become an important tourist attraction and with hundreds of thousands of visitors annually they are playing an important role in education on the need for conservation. Whilst these mostly specialise in tropical species, they inspire many people to want to do something for wild British butterflies in their "own patch".

Hopefully this book will encourage the reader — young or old — to actually start to create a garden area which will become increasingly attractive and useful to butterflies as it develops. Whether created in a home garden or the corner of a school field, it will become a constant source of interest and pleasure to the butterfly gardener.

2
Choosing the site

The size of the area you devote to the project will obviously depend on the size of your garden and the portion of it you are prepared to allocate for a butterfly garden. Before getting out the spade, some thinking and planning are in order!

There are basically **two types of butterfly garden**, although they can be combined to some degree if required.

A garden to attract butterflies : this first type provides flowering garden plants, perhaps in a border, specially selected as being useful to butterflies as a nectar source. If reasonably close to a wild area of countryside (a rough hedge or ditch, a patch of uncultivated ground, clumps of weeds and nettles, or fields and woodland), this type of garden is rather like a butterfly "pub" where the insects can fly in for refreshment.

A wild area for butterfly breeding : this second type of garden is where wild habitat is created containing the caterpillar foodplants so that the complete life cycle of certain species can be spent within the garden.

Many people will only have space for the first type, and may not anyway wish to let part of their garden go "wild". However, if there is room and it is practicable, the second type of butterfly garden is much preferable as it produces the environment in which courting, mating and egg-laying by adult butterflies can be observed. The earlier stages of egg, caterpillar and chrysalis can also be closely studied.

Shelter is very important to butterflies which tend to avoid exposed windy areas. I have often observed butterflies such as *Ringlets* happily flitting along the sides of steep ditches where the air is still, when the wind is howling around above ground level where there is not a butterfly to be seen. Walk round a wood on all but the stillest, sunniest day and the majority of butterflies will be seen in the warmest sheltered spots.

Therefore the plan must be to create this sort of favoured site in your own butterfly garden. Ideally, shelter could be afforded by a hedge or fence in the rear (north side) of the garden. However, if you live in an area where, say, easterly winds are common, some barrier against the effects of these would also be advantageous.

The site I chose for my own wildlife garden was quite exposed, so my first task was to plant a mixed deciduous **hedge** along the eastern side (against cool NE winds). This is predominantly *hawthorn,* but also contains *holly, honeysuckle, bramble, blackthorn* and *oak.* If you can plant a similarly mixed hedge, this will help more butterfly species find a home there.

On the north side I have made a 120cm **bank** of soil and turf. You may want to do some landscaping elsewhere in the garden, or even dig out a small pond, and a bank is both a good dump for the excess soil and a great help in sheltering those butterflies you hope to attract.

Starting a Butterfly Garden

I have recently worked on a large butterfly garden within the grounds of Lincoln castle. This site was chosen with the very high castle walls to the rear, providing a sheltered sun-trap, ideal for the purpose.

Try to site your garden with a **southerly** aspect to get as much sunshine as possible. Butterflies are usually only active when the sun shines and as soon as your flowers are put in the shade, the butterflies will fly off to seek a sunnier position elsewhere.

Of course, you may have existing fences, hedges, trees, or buildings which provide shelter already. Check the wind and sunshine in the garden over several days before finally deciding on the area that you will make into a butterfly garden.

What about the **size** of your butterfly garden? There are no limits here. A single buddleia bush will certainly attract butterflies, and a small clump of nettles or scrub will provide a habitat. A wider range of suitable flowering plants — and therefore of border space — will do a better job, giving you the opportunity of seeing a number of species. So, make up your mind after considering what else you need in your garden area (lawn, ornamental flower areas, pond, play area etc.). Remember that a butterfly border will need maintenance to keep it at its best. A wild patch of any size has implications for neighbouring areas — perhaps someone else's garden! Weeds and nettles may be a delight for the insects, but site them so as not to cause problems.

Butterfly Border

3
Butterfly borders

If you are going to use an existing border for your new butterfly garden, you should first consider removing any plant species which are not used by butterflies, such as *roses, gladioli, hydrangeas* and *peonies*. Usually the gaudy, large-flowered plants are unsuitable and the cultivars or modern strains of flowers often lack the basic scents attractive to butterflies. Butterflies generally like single "flat faced" flowers (such as *moon daisies*) enabling them to sit and feed without having to stick their proboscis or tongue very far in to extract the nectar.

Making a **new** border will be no different from creating any other patch of garden. Mark out the edge of the border, and dig the whole area. Systematically remove any plants and roots you find : removing couch grass now will save much time later. Rake roughly level, and then put in your choice of plants, working to a plan and being careful to allow the larger herbaceous plants and shrubs plenty of room for growth. Add some peat under the root ball and firm the soil around each plant as you go. Put in labels.

Butterflies may be expected to visit gardens from March to October or even November in a good year, so it is no use just having spring flowering plants in the border which would be flowerless for the rest of the summer and autumn.

What plants to choose and how to arrange them? **Appendix A** provides a quite comprehensive list of plants which are known to attract butterflies. This should be of great assistance in choosing which plants to have in your garden. An approximate indication of size, the usual colour and the flowering times are included to help plan your lay-out. No attempt has been made to separate garden and "wild" plants as this is often a matter of personal opinion.

Wild Area

Starting a Butterfly Garden

It is usually best to plant the larger species at the back of the border moving down to the smallest at the front. Colours and flowering times should be taken into account as with any other garden to obtain a pleasing appearance for human eyes as well as butterflies!

Some plants prefer different types of soil or climate and there are many good wildflower and gardening books (see Appendix C) that will give you this information e.g. *purging buckthorn* grows best on calcareous alkaline soils whereas *alder buckthorn* prefers acid peaty soils. Visit a local garden centre and talk over your requirements with the staff. Choose at least some plants which you see growing vigorously in nearby gardens, and try to avoid packing too many varieties into a limited area.

Once you have your flower border stocked with nectar sources attractive to butterflies, it will need managing in much the same way as a conventional flower garden. Cutting the dead heads off when the flowers are over will encourage further blooms on many plants. The usual tidying operations to prevent smaller and more delicate plants and shrubs from being crowded out by the larger dominant ones will be necessary. The fewer bare spaces between plants the better (though allowing for growth in the early years) and the aim should be an undulating mass of colour and scents over a long flowering period. Keep chemical sprays, insecticides and herbicides away from the flower border.

The **buddleia** or *butterfly bush* as it is commonly known, in all its varieties, grows rapidly and requires periodical pruning. Traditionally, this is done in the autumn after the last of the flowering plumes have withered. If you can put up with the bush unpruned until the spring and then cut back you will find that it flowers later. By having a series of buddleia bushes pruned at varying times from autumn to spring, you can provide a feast for butterflies until the first frosts of late autumn. If the uncut bushes look a bit untidy during the winter, remove just the dead flower heads to improve the appearance and also help to prevent uprooting in high winds — these bushes tend to have shallow roots and become a bit top heavy.

There are many records of *spotted flycatchers* learning to "stake out" *buddleia* bushes, systematically harvesting the butterfly visitors, so it is not advisable to encourage these birds by the provision of unsuitably sited nest boxes!

Buddleia **cuttings** can be taken in autumn or spring and are usually successful if kept moist and planted in sandy soil. The cut should be made straight across just below a joint. Cuttings should be 15 to 30cm long with leaves from the lower third removed (the part of the stick to be inserted in the ground). After planting in the required position, regularly inspect them to make sure they have not been dislodged by the wind.

Apart from the plants, there are other ways of attracting butterflies into your garden. **Rotting fruit** and **sap** from a wounded or dying tree attracts the *Red Admiral* (and also wasps) in Autumn, **Aphid secretion** on leaves of ash and other trees attracts the *Purple Hairstreak* and the *White-letter Hairstreak* in August.

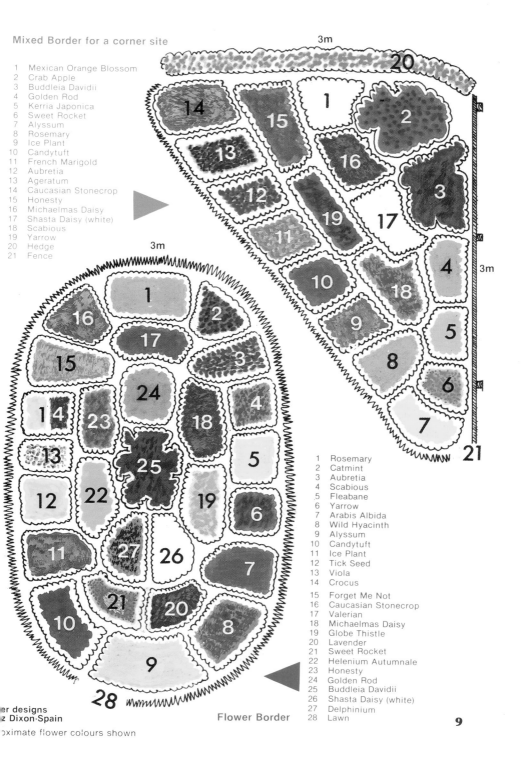

Mixed Border for a corner site

1 Mexican Orange Blossom
2 Crab Apple
3 Buddleia Davidii
4 Golden Rod
5 Kerria Japonica
6 Sweet Rocket
7 Alyssum
8 Rosemary
9 Ice Plant
10 Candytuft
11 French Marigold
12 Aubretia
13 Ageratum
14 Caucasian Stonecrop
15 Honesty
16 Michaelmas Daisy
17 Shasta Daisy (white)
18 Scabious
19 Yarrow
20 Hedge
21 Fence

1 Rosemary
2 Catmint
3 Aubretia
4 Scabious
5 Fleabane
6 Yarrow
7 Arabis Albida
8 Wild Hyacinth
9 Alyssum
10 Candytuft
11 Ice Plant
12 Tick Seed
13 Viola
14 Crocus

15 Forget Me Not
16 Caucasian Stonecrop
17 Valerian
18 Michaelmas Daisy
19 Globe Thistle
20 Lavender
21 Sweet Rocket
22 Helenium Autumnale
23 Honesty
24 Golden Rod
25 Buddleia Davidii
26 Shasta Daisy (white)
27 Delphinium
28 Lawn

er designs
z Dixon-Spain

Flower Border

oximate flower colours shown

9

4
Wild area for butterflies

A herbaceous border may be quite enough to cope with, but if you wish to be more ambitious and create a **mini-meadow** out of, say, an existing piece of lawn you will need to remove the turf and the first 5cm of topsoil from the site. Don't worry about losing the goodness from the ground as most wildflowers prefer poor soils! It is no use scattering wildflower seeds on top of existing grasslands as the plants will be unable to find a niche in which to grow. Small bare patches of soil can be made in existing lawns. Either sow seed directly or use seedlings you have grown in a seed tray.

My own garden is close to a dismantled railway line where 25 butterfly species have been recorded. The bottom third of the garden, nearest to the old track, is given over to wildlife and particularly butterflies, whilst the flower borders near the house are stocked with favoured garden nectar plants. I like to keep all the native wildflowers and grasses separate from the exotic cultivated species normally accepted as garden plants, but you may prefer to mix them, particularly if the available area is quite small.

Prepare the wild patch, then, by removing topsoil and weeds/grasses and then work to a plan as for a border. **Appendix B** provides a detailed list of plants and grasses which are used for egg laying and caterpillar feeding. Many are available as seeds and **Appendix D** gives details of how to obtain these. Others can be introduced as cuttings from existing plants, or bought from garden centres. Your own choice of plant-mix will depend on many factors, but we will now review the main possibilities for butterfly breeding and feeding locations.

Nettle Patches
As *stinging nettles* spread vigorously, it may be as well, in a small garden, to restrict the spread of roots by confining them to an old tub or bath or the like. *Small Tortoiseshells* and the other nettle feeders always select new succulent growth for breeding purposes, so provide this at the time of year when eggs are likely to be deposited.

Divide your nettle patch roughly in half and scythe down to ground level the southernmost part at the beginning of July. Any spring broods of caterpillars will have already utilised the nettles so take care that chrysalises are not attached. The taller nettles at the rear are possible pupation sites while the re-growth at the front is for the next generation of butterflies to lay on.

The whole patch of nettles should be scythed down in late autumn as all nettle feeders will then be in their adult form, away from the nettles, feeding up and selecting hibernation sites for winter. If you happen accidentally to cut down a stem with a "nest" of caterpillars, carefully place it amongst the uncut-ones and the tribe will soon make their way on to the living foodplant. The caterpillars fall off very easily, so place a bowl underneath when you are transferring the cut stem.

Grassland

Probably the best service the butterfly gardener can render is to provide the natural grassy habitats in which many of our native butterflies reproduce. The minimum useful area of grassland will be about three square metres, in a warm, sunny and sheltered position. Butterfly species need different heights of grass sward. A mowing or scything regime should be worked out with this in mind, and the sections divided by regularly mowing short grass paths. Even these will be used for basking, courting and territorial displays by species such as the *Wall Brown,* and the paths will give you access to the whole area without having to wade through taller grasses and flowers.

It is best to sow **grass and wildflower seed mixture** in late October, for flowering the following year — if sown in spring you are less likely to get flowers the same year. Mix the seeds with damp sand and scatter sparsely onto your prepared ground by hand. Lightly rake over and water in dry weather to help germination. If birds are going to be a problem eating the seed, get the children to make a scarecrow! There are several types of grass/flower seed mixes. Choose one which suits your local soil type and best fits your own ideas for species. If you want a wild flower not in the mixture, buy this separately and mix in before sowing.

Once you have an area sown with a good selection of native grasses, different species will flourish with the cutting cycles. Remember though that species need to be present locally before they can colonise **your** wild area.

If you have created a mini-meadow by sowing patches of wild flowers, these should be catered for in the system so as not to cut down favoured ones until after they have flowered and set their seed. The grass should be cut in sections of differing heights, as just described.

In my own garden, I have a **spring** mini-meadow area which extends into the woodland area. I cut this in March and then leave it until late July when several further cuts are made through till winter, not letting the grass grow more than about 20cm tall. My **summer** mini-meadow area is only given a late scythe at the end of October.

There are so many options it is difficult to give general advice on when to cut, but aim at creating a mosaic of various heights of grass. This is best achieved by never cutting the whole area at one time (except perhaps in the winter). Never add fertiliser. Leave the cut grass a day to allow any caterpillars to crawl away and then rake off and remove.

Do not be afraid to be selective in sparing flowers and plants you wish to encourage. This is part of the art of wildlife gardening. Your practical knowledge of the wild plant species that thrive in different circumstances will soon build up. There are so many variables on soil type and condition, regions, weather in a particular year and so on to take into account. Show off your garden to interested visitors and you will no doubt receive odd bits of valuable advice to use in future years.

Annual cornfield
A patch of ground can be sown with several wildflowers and perhaps grasses (or even wheat!) cut down each late summer, and then dug over in the autumn. The seeds should then germinate in the soil and grow again the following spring. This is really mimicking the old cornfield cycle before the attractive "weeds" were sprayed out of existence. Packs of cornfield seeds are available (see **Appendix D**) thanks to the growing interests in these beautiful plants which used to be so common. Note that you **must** turn over the soil each autumn — these plants thrive on disturbance, in complete contrast to the wild flowers of the grass meadows.

Hedges
A surprising number of butterflies make use of hedges with species such as the *Holly Blue* flitting over them to lay their eggs on *ivy* buds, and many other species such as *Gatekeeper, Red Admiral* and *Speckled Wood* resting on the leaves to sun themselves, or going there to roost for the night. Brimstones hibernate within clumps of *ivy* or *bramble* which should be encouraged to intertwine with the other shrubs and bushes making up the hedge. The greater the diversity of these, the more interesting a feature your hedgerow will become. Prune the various bushes at different times and let them grow as tall as circumstances will allow (eg. the tolerance of your neighbour!) The taller and thicker the hedge, the more effective the windbreak and sun-trap. If space permits, let some bushes grow forwards, to produce a scalloped effect along the hedgerow. The best wind-break is from a hedge with an "A" cross-section.

Although a deciduous hedge of *hawthorn, blackthorn, privet, holly, honeysuckle, buckthorn, sallow* etc., is much preferable, a thick screen of quick growing *leylandii* hedging can also produce a very effective sheltered area. The tops should be lopped when they have reached the required height. Then prune back each April and August.

Trees
Take care not to plant too many trees in a small area, and keep in mind the eventual size they will attain. Too much **shade** may make your garden useless as a butterfly sanctuary. Try to let the native *oaks, elm, holly etc.*, grow to their natural size without pruning. Lower branches should be left as these are often utilized by *Hairstreaks* for egg laying.

Brambles
These can be cut back annually to keep the bushes reasonably under control for they spread vigorously if left unchecked.

General
Try not to be too tidy. A bit of neglect really can work wonders. A pile of old logs can provide a hibernating site for *Peacocks*. Wildflowers and grass seeds in individual packets and various mixes are available from a number of seedsmen but there is little harm in collecting ripe seed from **common** flowers and grasses in the countryside. On no account attempt to take seed from rare wild flowers and if in doubt leave them alone.

When you get home you can either introduce the seeds into your mini-meadow or cultivate them in trays and pots to plant out as seedlings or young plants. The seeds of several species need to overwinter or be scarified (rubbed/scratched) before they will grow. **Never** dig up plants growing in the wild. As well as being morally wrong, it is now illegal. Most of them are now protected under the Wildlife and Countryside Act 1981.

You probably don't have room for all the habitats just described, but you should now be able to make your choice and get started on your wild area.

5
Gardens and butterflies

Strictly speaking there are no garden butterflies, but certain species are much more likely to turn up and even breed in your butterfly garden than other rarer or more specialised ones. It would probably be a pointless exercise to plant *broad-leaved sallows* in the hope of enticing a *Purple Emperor* to visit. However, if your garden happens to be next to a very large old oak woodland in central or southern England, it might be worthwhile. I have read of a resident in such a situation who occasionally used to have these beautiful insects resting on damp washing hanging out on the line! I have had *Red Admirals* do the very same thing in my own garden. Last year a friend identified a *Camberwell Beauty* — a rare immigrant from Scandinavia — on a neighbour's washing.

So nothing can be discounted but do not **expect** to have any but the more usual types. Luckily, these are just as attractive and appealing as the rarities. There are now several good butterfly identification books available (see **Appendix C**) and though it is not possible in this book to go into great detail about the 60 odd species on the British list, we can at least give you information about the common species.

I will concentrate on 23 butterflies likely to visit your garden for nectar and/or breeding. These are in the main the more common butterflies, less particular in their habitat requirements. Many are living on the northern edge of their range in Britain, so bear in mind that not all will be found in your part of the country. For example, the *Gatekeeper,* whilst common over most of England, is not found much further north than South Yorkshire.If you do live near to a specialised butterfly habitat you may be able to persuade that species to inhabit your garden. At How Hill in Norfolk, *Swallowtails* regularly fly in the gardens not far from their breeding grounds in the nearby Fens and Broads.

The Nymphalidae
The most striking aristocrats of the butterfly world to be commonly seen in gardens belong to the **Nymphalidae** family. Of the five members to be dealt with here, four breed on *nettles* — the *Small Tortoiseshell, Peacock, Red Admiral* and *Comma;* while the *Painted Lady* prefers thistles. You may have to put up with clumps of nettles in order to see these attractive butterflies flying around your flower borders later on.

Many people think they are doing the correct thing by leaving a nettle patch in an out-of-the-way corner stuck away behind a fence or next to the compost heap, little realising that they are only providing half the needs and wondering why they never see webs of tiny *Tortoiseshell* or *Peacock* caterpillars. The position in which the nettles grow is the often forgotten factor. The female *Tortoiseshell* will always select young tender nettles growing in full sunlight in a situation well sheltered from the wind. This may be on nettles growing in a natural hollow in the ground, or in front of a bush or a fence to the north, or on a south-facing bank.

After first feeding up on nectar to replace the energy lost during the long winter hibernation, the adults court and mate. The female then lays her eggs in one or more large batches on the underside of the nettle leaves towards the top of the plant. When the caterpillars hatch, after a couple of weeks, they spin a web as a protection against birds and insect predators and live gregariously in this until June, now fully grown, they wander off to change into a chrysalis. They emerge as butterflies in another fortnight and live only for a few weeks.

The cycle is repeated with the second brood of adults emerging in August. These butterflies spend the rest of the sunny days of summer and autumn feeding up before their winter sleep. They often choose garages or sheds for their hibernation roosts, so it is worth leaving windows or doors open in late autumn. Don't forget to let them out again in Spring as many butterflies perish each year through not being able to escape back into the garden.

Peacocks have a similar life cycle, but only produce one brood. They lay their eggs later in the spring and generally hibernate earlier — in more outdoor sites, such as hollow trees or piles of wood.

Commas usually spend the winter perched on a tree trunk exposed to the elements. After mating in April, the female lays her eggs singly on the under-side of nettle leaves. Like the *Small Tortoiseshell* and *Peacock*, the *Comma* will breed in gardens, but it is not seen in such large numbers and its geographical range, though steadily expanding in recent years, is nevertheless restricted to the country south of the Humber.

Red Admirals, seldom withstand the cold of our winters. From early summer onwards they migrate here from the continent. Eggs are again laid singly on nettle leaves — top sides this time. Caterpillars construct conspicuous tents out of the leaves by silking up the edges. In late autumn this butterfly is very fond of rotting fruit and a pile of over ripe plums positioned on the warm side of a compost heap should provide an irresistible temptation.

The *Painted Lady* is also an immigrant but not usually so common as the *Red Admiral.* Female *Painted Ladies* breed on thistles. They usually perish when the first frosts arrive.

These five species are all very attractive large butterflies, often seen together in gardens in large numbers late in the year feeding on flowers such as *buddleias, michaelmas daisies* and *ice plants.*

Red Admiral

Brimstone

Peacock

Holly Blu

Small
Tortoiseshell

Purple
Hairstreak

Comma

Gatekeeper

Not to Scale

Painted Lady

17

The Browns
Seven species of the **Brown** family regularly turn up in gardens. They all breed on different grasses and so if you make a mini-meadow area you should be able to provide a home for some of them. The larger the meadow area, the greater your chances of success.

Chocolate coloured *Ringlets* favour damp grassy areas, while *Small Heaths* are usually found in drier situations. The *Marbled Whites,* with very pretty black and white chequered markings, are mostly confined to the South of England. They are very partial to the flowers of *scabious, knapweed* and *thistles.* Instead of gluing her eggs to a blade of grass as the others do, the *Marbled White* female will fly slowly over suitable grasses and "bomb" her eggs at random.

These Browns all live in colonies, some being rather more loose-knit than others. *Gatekeepers* seem to flit for endless hours over relatively small bushy territories. Their favourite flowers are *bramble,* but *marjoram* is well liked in gardens. *Wall Brown* males are very territorial and will stake out a home "base" on a patch of short grass awaiting virgin females with which to mate or rival males with which to spar. Its close relative the *Speckled Wood* has much the same habits, but prefers woody areas of dappled sunlight, so you really need plenty of trees in your garden to attract this butterfly.

The *Meadow Brown* is one of our commonest butterflies and has rather a lazy, feeble flight. Like the *Ringlet,* it will fly even in the dullest weather and is also a frequent visitor to *knapweed* and *thistle* flowers.

The Skippers
The **Skipper** family includes three butterfly species likely to be encountered in gardens. *Small Skippers* have a swift flight and are more likely to be seen at rest on a grass head, or feeding on favoured flowers such as *knapweed.* In sunshine, they bask in a typical pose, with forewings half open and hind-wings fully open and flat. They fly within small colonies in July and August. Their eggs are laid in rows inside the sheath of a soft grass such as Yorkshire fog.

The taller patches of grassland in your wild area will suit **Skippers** best. **Small Skipper** caterpillars spend the winter in a sheath on a grass stem and the chrysalis is also made inside a cocoon spun amongst grasses.

The *Essex Skipper* is very similar but more local, being found only in south and south-east England. It spends the winter as an egg, again inside a grass sheath. It also lives in colonies often alongside its cousins, the *Small Skippers,* but generally appears a week or so after, and lasts that much longer.

The *Large Skipper* is more solitary, but its flight is even swifter and stronger than its smaller namesake. It makes an appearance before the other two and is usually on the wing early in June. Once again most of its life cycle is centred on grasses where the eggs are laid and on which the caterpillars feed, changing their skins several times before constructing little hides into which they crawl to spend the cold winter months. These winter quarters are called the hibernaculum and are constructed by fastening together several grass blades with silk. The caterpillar emerges again in the following March to complete its growth, which will have taken ten or eleven months.

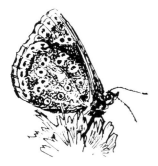

The Blues and Coppers

Old established gardens with *ivy* covered walls and *holly trees* may be the home of the *Holly Blue* butterfly, a double brooded species, which lays its eggs on holly in May and the next generation on ivy in August. It may be found fluttering over bushes, hedges and trees. The underside of the wings is a refreshing icy blue.

In June, its close relative, the *Common Blue* is on the wing. These keep much closer to the ground breeding and feeding mainly on *bird's foot trefoil,* growing in sunny sheltered positions. They live in colonies and a second brood of adults emerges in August. There can be few more delightful sights in summer than seeing blue butterflies flitting over the yellows and reds of the *bird's foot trefoil.* If you plant several clumps of the food plant, you may see them too!

Growing plants in large patches is an important point. I have never seen reference to butterflies being attracted to daffodils, yet every year the *Small Tortoiseshells* and *Peacocks* emerging from hibernation make for these flowers growing in great sheets in one Lincoln park. These are often the first butterflies I see. *Spring Peacocks* also make for the *bluebells* carpeting the woodland floor though I have never seen any on individual daffodils or bluebells in a garden setting.

The *Small Copper,* a fiery little butterfly, is quite a pugnacious character, especially the male when defending its territory against other insects. It lives in small colonies and occurs all over Britain, so you have a good chance of attracting this pretty butterfly into your garden. It is quite fond of feeding on *ragwort* flowers, but if you grow them take care as they are poisonous to livestock.

Starting a Butterfly Garden

The Whites

The only British butterflies that do any harm to man are the cabbage eating *Large* and *Small Whites.* However, even these have interesting life cycles and who can deny that summer would not seen the same without their happy, carefree flights around our gardens.

If you do want to grow *cabbages* and other **brassicas** in a vegetable patch, as well as garden for butterflies, what can you do? Well, *Large Whites* also breed on *nasturtiums,* whilst *Small White* caterpillars will eat *aubretia,* so if you discover cabbage or brussel sprout leaves with batches of colourful *Large White* or individual green *Small White* caterpillars, why not just pick off the infected leaves and place them amongst the alternative food plants. This is easier than spending time and money on chemical sprays that, as well as killing the caterpillars, will also do away with "innocent" creatures, and may do yourself no good into the bargain.

The *Green-veined White* (and for that matter the *Orange Tip* female) looks very similar to the *Small White* at a glance. This delightful butterfly has no interest whatsoever in the cabbage family, preferring *ladies smock, hedge mustard* or *garlic mustard* to deposit her eggs. They are attached to the underside of the lower leaves.

The *Orange Tip* may be seen any time from the beginning of May until the end of June. The more conspicuous males, who alone have the orange tips to the forewings, emerge well before the less noticeable females. *Orange Tip* males patrol up and down their favourite habitat investigating any white object in their search for a mate. The butterflies often rest on the flowerhead of a *cow parsley* where the motley greenish-yellow patterning on the underside hindwings make them very difficult to spot.

Their preferred foodplants are *ladies smock* (which likes damp places and would be a very pretty addition to your meadow area, if part of it stays a bit on the wet side) and *garlic mustard* which likes drier sites, and seeds itself very readily. If you can entice this pretty butterfly to its caterpillar foodplants you can almost guarantee to find the tiny bright orange bottle-shaped eggs on the flower stem just below the florets. Often several eggs are deposited on a well sited plant, in a sheltered but sunny position. After about a week, the first caterpillar to hatch will show its canibalistic tendencies by devouring any other eggs or small caterpillars it comes across. This is nature's way of ensuring at least one of the offspring has sufficient food to reach maturity. The chrysalis is shaped just like a large brown thorn and is attached to a nearby stem. In this guise the *Orange Tip* spends the winter — another reason for leaving rough areas of long grass in your wild patch.

The *Brimstone* is always a welcome sight in the spring, being one of the first butterflies to be seen. The male is a richer yellow than the female and the very word butterfly is supposed to have originated from the old countryman's description of the **butter-coloured fly.** The males wake first from their hibernation and seek out the *buckthorn* bushes, where the females will congregate later to mate and lay their eggs. *Alder* and *purging buckthorn* are the sole foodplant and are often scarce, but the butterflies show an amazing sense of location in finding these bushes. You can help the species to survive by planting one or two of these bushes in your garden.

The new generation of male and female butterflies emerge together in July. They show a rather strange and unique taste for the red *runner bean* flowers. I once watched a pair spend a whole afternoon on these flowers. Occasionally, one would briefly fly off, but it would soon be back on its preferred flowers on which it was obviously "hooked".

The Hairstreaks

If you have mature *wych elms* or *oaks* near your garden there may already by a colony of *White-letter Hairstreaks* or *Purple Hairstreaks* living in the canopy. Occasionally both species descend to feed on flowers such as *creeping thistle* and *bramble.* A relative of mine has a large oak at the bottom of his garden. On a sunny August evening you can sit on his lawn and watch *Purple Hairstreaks* spinning like silver coins through the uppermost foliage. The effect is produced by the silver and purple wing markings glinting in the sunlight. This garden is on the outskirts of a city, quite a distance from the nearest oak wood — the usual habitat of this butterfly.

The *White-letter Hairstreak* has suffered in this country indirectly with the ravages of Dutch Elm Disease. A disease resistant strain of *Elm* (Sapporo autumn gold) has been developed which appears to be immune from attack. If you are unfortunate enough to have a dying elm, you might save the butterfly colony by planting one of these hybrids nearby. By the time the tree actually dies, which usually takes a number of years, these quick growing saplings should be large enough to provide egg laying sites. (See **Appendix D** for a supplier).

6
Butterflies — notes on natural history

Tempting butterflies into your garden — and perhaps getting them to breed there — may be a little easier if you know some details of their natural history. So we will look briefly at a few aspects, and hope you will delve deeper, perhaps through books like those listed in **Appendix C.**

Butterfly life-cycle
1. Mating
Exciting courtship flights of a male and female can often be seen, as a preparation to mating, which may take from 1 to 10 hours. During mating, the pair can still fly — important to escape predators.
2. Egg laying
In spring and summer you may see females laying eggs on the foodplant of the caterpillar. The eggs are usually placed on the underside of leaves or on the stem, out of direct sunlight and away from rain. They may be laid singly, or in group up to 400 eggs, of diameter 1mm to 4mm. The eggs are spheres or cylinders.

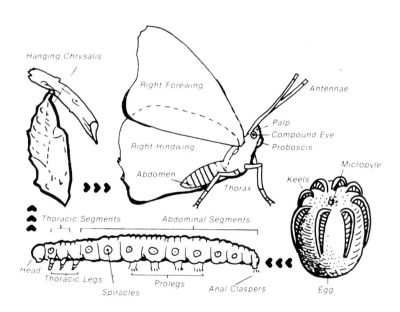

3. The caterpillar
The emerging caterpillars (or **larvae**) eat a hole through the tiny egg shell, then eat the remains of the egg. It won't be hard to see the effects of caterpillars on their food plants, and as they grow (eating young leaves for preference) they change skins perhaps 5 times before reaching full size.

4. The chrysalis

Caterpillars find a suitable place on the food plant and then spin a small silk pad and attach their tail. They then hang upside down and shed their last skin, becoming a chrysalis (or **pupa**). This is vulnerable to predators and so is usually well camouflaged. Little can be seen on the surface, but inside a marvellous change is taking place, over several weeks or indeed over the winter months.

Watching the emergence of an adult from the chrysalis is fascinating — the complex wings seem impossible to fit inside the tiny sack. Within an hour or so the adult flies off, seeking nectar. Most live for only a few weeks.

6. Hibernation

All UK resident butterflies hibernate but the stage of the life-cycle varies with the species. Some have two generations per year, and some just one. Tropical butterflies may have up to 12 generations in a year.

Page 24 shows details life-cycles of four common species.

What do butterflies see?

Although we cannot be certain what butterflies see, we do know that they can detect **ultra violet** light, and therefore see patterns on flowers which are invisible to us, unless we shine a u/v source on them. Two seemingly similar flowers may then look very different.

This may be one reason why some flowers attract specific butterfly breeds. For example, runner bean flowers only seem to lure Brimstones. Perhaps butterflies are attracted to particular colours during the day (purple is an obvious favourite), or is it that some flowers give off stronger perfumes at certain times during the day and indeed their flowering period?

There is still much to be learnt about butterfly preferences and the reasons for them - and plenty of scope for the amateur investigator here, at home or at school. All you need is patience. Watch and make careful notes of time, weather, flower, butterfly type, behaviour. Does the same flower-type get equal attention in different parts of your garden? Does the amount of shelter, and the level of shade from the sun, affect where the butterflies feed?

Your local BBCS branch will be delighted to receive your findings - many of them run butterfly garden schemes and publish regular newsletters which include keeping members informed as to what others have seen or discovered.

Enemies

Butterflies are eaten by birds, spiders and insects. They also suffer from our own activities. Man cuts down forests and hedges, sprays chemicals on crops and roadside verges, and changes farming techniques. Butterflies camouflage to try to protect themselves against predators, but have little chance against man. We can help by setting up butterfly gardens and wild areas, and also by sparing hedges and roadside verges from cutting back. Most of all, we can help by involving children in conservation, so they understand the needs of the other species which share the planet.

Starting a Butterfly Garden

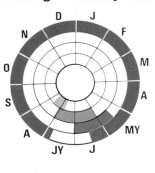

Brimstone

- ■ BUTTERFLY
- ■ EGG
- ■ CATERPILLAR
- ■ CHRYSALIS

Habitat
Open woodland or rough bushy land where its foodplants grow.

Life cycle
One generation a year. Over-winters as a butterfly.

Painted Lady

- ■ BUTTERFLY
- ■ EGG
- ■ CATERPILLAR
- ■ CHRYSALIS

Habitat
Rough ground, hillsides, and lanes.

Life cycle
One to two generations a year. It cannot survive our winter in any stage, being killed off by the arrival of the cold weather at whatever stage has been reached.

Small Tortoiseshell

- ■ BUTTERFLY
- ■ EGG
- ■ CATERPILLAR
- ■ CHRYSALIS

Habitat
Almost any type of country.

Life cycle
Two generations a year. Over-winters as a butterfly.

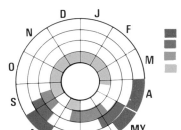

Holly Blue

- ■ BUTTERFLY
- ■ EGG
- ■ CATERPILLAR
- ■ CHRYSALIS

Habitat
Woods, shrubby areas, hedge-rows, and gardens.

Life cycle
Two generations a year. Over-winters as a chrysalis.

Observe and record
In the previous chapter, we mentioned that you might use your new butterfly garden to study the habits of the resident and visiting butterflies. This can be extended to include a count of species — note down dates and numbers seen. Schools, in particular, will want to use their gardens and wild areas for such studies. Curiosity and careful observation are excellent starting points for science.

Use as a stimulus
You may prefer to photograph or sketch the insects, and many schools will use the butterflies as a stimulus to topic work encompassing flight, movements, insect life-cycles, measurement, symmetry, colour, and lots of expressive drama and writing.

A butterfly greenhouse
If you are a school without suitable grassy areas you may instead create a **greenhouse** for rearing butterflies. Use the standard glass variety, or make your own from metal and plastic sheet, and be sure the door and vents are covered by netting to keep hatched butterflies inside. **Research** the butterfly types you want. The **appendices** will help here. You need a source of caterpillars or chrysalides (local gardens or bought commercially — see **Appendix D**) and the correct food plant, growing inside the greenhouse. Put the two together and watch! When adults appear, provide nectar (20% honey, 80% water) in shallow saucers, perhaps with rotting fruit for landing pads, to provide food. Place a blue, red or purple circular **scouring pad** in the saucer to attract the butterflies. Some prefer real flowers, so grow pots of *busy lizzies* or *lantana* in the greenhouse.

Don't try to breed too many butterflies, or too many varieties. You will probably want to let the adults fly off to breed, after about a week in the greenhouse. With encouragement, the children will respect the butterflies as wild insects to be marvelled at. Watch out that sun and heat does not dry out your greenhouse — ventilate by opening the door and roof vent (covering with netting) and shade the roof with paint or plastic from a garden centre.

Caterpillars
Children can usually find caterpillars, on stems of their food plants, and these can be the basis of interesting experimental work. Remember of course that the caterpillars should be returned to the garden (and their food plant) at the end of the day, and treated with respect by the children.

Try shielding half of a stem from light (with black paper). Do the caterpillars choose light or shade (or both)?

Wet a stem (with repeated showers from a fine spray) and watch how the caterpillars react. Compare with caterpillars on a "dry" stem.

You will think of other possible experiments; and weighing, measuring and drawing activities are further ways of getting children actively involved with the insects.

Starting a Butterfly Garden

Breeding butterflies
One way of increasing the number of butterflies in your garden is to try and rear one or two types for eventual release. This is a controversial subject but rearing does allow you to observe closely the wonder of the butterfly's complex life.

First you need the eggs or young caterpillars of a common species such as the Peacock butterfly (see **Appendix D** for a supplier). Do not collect or purchase too many to start with, as rearing butterflies can be quite time consuming.

The secret of successful breeding is to keep the livestock in as near their natural conditions as possible. Any attempt at 'molly coddling' will end in disappointment. For example, many people might be tempted to put the eggs or chrysalides in a warm place to hatch (eg. an airing cupboard) where they will probably die of desiccation.

Peacocks breed on stinging nettles so a young clump should be dug up and planted in a 15cm pot some time beforehand. "Fresh cut" nettles are no substitute as they quickly dry out and are little use as food. Young caterpillars can be put straight on to the nettle leaves, but eggs should be left in a small clear plastic box until after they have hatched. Caterpillars can then be transferred with a small paint brush.

The pot and nettles should be covered with fine black netting, held in position by supporting wires bent over with their ends in the soil (see diagram). The bottom of the netting is tied round the rim of the pot with string.

The caterpillars will change their skins several times and form a chrysalis within a few weeks. This may be attached to the food-plant or the netting. Keep the pot out of direct sunlight. The nettle plant will need regular watering. The adult Peacocks should emerge in all their glory after another couple of weeks.

If you wish to try to get adult butterflies to pair and lay eggs, it is beyond the scope of this little book to do more than recommend further reading on this subject (see **Appendix C**).

Breeding Cage

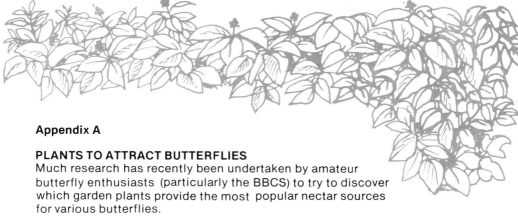

Appendix A

PLANTS TO ATTRACT BUTTERFLIES

Much research has recently been undertaken by amateur butterfly enthusiasts (particularly the BBCS) to try to discover which garden plants provide the most popular nectar sources for various butterflies.

Although no list is ever complete, the following is a reasonably comprehensive guide. Plants are arranged according to size. After the name, flower code and flowering months are given, in abbreviated form. **Wild** and **garden** plants are included.

Small plants

Ageratum	blue June-Aug
Alyssum	white June-Aug
Anaphalis margaritacea	pale white Jul-Aug
Apple mint	pink Aug-Sep
Arabis albida	pink/purple Mar-May
Aubretia	red/mauve/blue Apr-Jun
Bergenia crassifolia	pink Mar-Apr
Bird's foot trefoil	yellow May-Sep
Bluebell	blue May
Bugle	dark blue May-Jul
Candytuft •	white/pink/mauve Jul-Aug
Caucasian stonecrop	pink/red Jul-Aug
Coltsfoot	yellow Apr
Crocus	various Mar-Apr
Daisy	white Feb-Nov
Dandelion	yellow Mar-Sep
Dianthus deltoides	pink May-Aug
Forget-me-not	blue Apr-Jun
Ground Ivy	purple Mar-May
Lobelia	blue/pink/white Jun-Aug
Lucerne	pale purple Jun
Lysimachia clethroides	white Jul-Sep
Marjoram	pink Jul-Sep
Phyopsis stylosa	deep pink Jun-Jul
Primulas/polyanthus	various Apr-May
Primrose	yellow Apr
Red clover	red May-Sep
Thrift	pink Apr-Aug
Thyme	pink/purple May-Aug
Tick seed	yellow Jul-Sep
Tormentil	yellow May-Oct
Viola/pansy	various Apr-Oct
Wild hyacinth	blue Feb-Apr
Yarrow	pink/white/red Jun-Oct

Medium sized plants

Agrostemma milas	pink Jun-Aug
Anchusa azurea	blue Jun-Aug
Betony	pink Jun-Sep
Catmint	mauve/yellow Jun-Sep
Cornflower	blue Jun-Aug
Daffodil	yellow Apr-May
Delphinium	blue/pink/white Jun-Jul
Erigeron species	blue/pink/yellow Jun-Aug
Erysimum linifolium	purple Jun-Oct
Fleabane	yellow Aug-Sep
Globe thistle	grey blue Jun-Aug

Medium sized plants — continued

Heliotrope	purple Jun-Sep
Hemp agrimony	pink Jul-Sep
Honesty	pink/white May-Jun
Hyssop	deep pink/blue May-Aug
Ice plant	pink Sep
(Sedum spectabile)	
Jasione perennis	mauve Jun-Jul
Knapweed, brown	pink Jun-Sep
Knapweed, greater	purple Jul-Aug
Liatris spicata	deep pink Sep-Oct
Ligularia species	yellow July
Marguerite	white/yellow Jun-Aug
Marigolds	various May-Sep
(African & French)	
Michaelmas daisy	white/pink/mauve Aug-Oct
(and other Asters)	
Mignonette	pale green/yellow May-Aug
Mint	purple Aug
Phlox drummondii	various Jun-Aug
Phlox paniculata	purple Jun-Aug
Purple loose strife	red/purple Jun-Aug
Ragged robin	pink May-Jun

Round headed rampion	blue Jun-Aug
Rudbeckia purpurea	various Aug-Sep
Sainfoin	red May-Aug
Saw-wort	pink Jul-Sep
Scabious, butterfly blue	blue May-Oct
Scabious, Devil's bit	dark blue Aug-Sep
Scabious, field	blue Jul-Aug
Scabious, small	blue Jul-Aug
Scabious, sweet	various Jul-Sep
Shasta Daisy	white Jul
Soapwort	pale pink Jul-Sep
Sweet rocket	pink May-Jul
Thistles	pink shades Jun-Sep
Toadflax, pale	pale mauve Jul-Aug
Valerian	red/white/pink Jun-Aug
Verbinas	various Jun-Aug
Wallflowers	various Apr-Jun
White dead-nettle	white Apr-Oct

Larger plants

Angelica	white Jul-Aug
Chrysanthemums	various Aug-Oct
Dahlia	various Jun-Sep
(Coltness hybrids)	
Golden Rod	golden Jun-Oct
Helenium autumnale	various Sep-Oct
Helichrysum	various Aug-Sep
Ivy (climber)	pale green Sep-Nov
Rose-bay willow herb	pink Jul-Sep
Runner bean	red Jun-Aug
Sweet William	pink/red Jun-Jul
Teasel	mauve Jul-Aug
Travellers Joy (climber)	white Jul-Aug

Shrubs and Bushes

Abelia schumanii	pink Jun-Sep
Blackthorn	white Apr
Brambles	white Jul-Aug (as well as blossoms, ripe and over ripe blackberries attract many butterflies).

Buddleias:

Davidii	purple/white Jul-Aug
Alternifolia	mauve Jun
Colvillei	pink Jun-Jul
Crispa	purple Jul-Aug
Fallowiana	purple/white Aug
Globosa	orange May-Jun
Lochinch	white Aug
Weyerana	yellow/purple May-Jun
Caryopteris x clandonensis	bright blue Aug-Nov
Caenothus x delilianus esp. 'Gloire de Versailles'	lt. blue Jun-Oct
Ceratostigma Willmottianum	bril blue Jul-Aug
Clematis heracleifolia	purple Aug-Sep
Clerodendrum bungei	pink Sep-Oct
Colletia armata	white Sep-Oct
Daphne odora (evergn.)	deep pink Mar-Apr
Escallonia bifida	white Aug-Sep
Escallonia macrantha	pink Jun-Sep
Heathers	purple/pink various
Hebe albicans (evergn.)	white Jun
Hebe x andersonii var. (evergn.)	blue Jun-Oct
Hebe brachysiphon	white Jun-Jul
Hebe x fransiscana (evergn.)	purple May-Oct
Hebe 'Great Orme'	pink Jun-Oct
Hebe 'Midsummer Beauty'	pale purple Jun-Aug
Hebe salcifolia	pale purple Jun-Aug
Kerria japonica	yellow Apr-May
Lavender 'munstead dwarf'	lavender Jul-Sep
Mexican orange blossom	white May-Aug
Privet	white Jun-Jul
Rosemary	mauve May-Jul
Sallows/willows	yellow catkins Mar-Apr
Senecio laxifolius (evergn.)	yellow Jun-Jul
Spiraea x bumalda	red Jul-Aug
Snowberry	pale pink Jun-Sep

Trees

Bird Cherry	white May
Crab Apple	pink/red Apr-May
Wayfaring Tree	white May
Wild Cherry	white May
Wild Pear	white Apr-May

Appendix B

PLANTS FOR EGG LAYING AND CATERPILLAR FEEDING

These plants (mainly wild but several having garden varieties)
provide "homes" for eggs and caterpillars. Butterflies which make use of them are also named.
The list is not complete — research continues!

Alder buckthorn	Brimstone
Bird's foot trefoil	Common blue, Dingy Skipper*, Green Hairstreak*, Silver-studdied Blue*, Wood White*, Chalkhill Blue*
Bilberry	Green Hairstreak*
Blackthorn	Brown Hairstreak*, Black Hairstreak*
Cabbage family	Large White, Small White
Charlock	Green-veined White, Orange Tip
Cinquefoil	Grizzled Skipper*
Common storksbill	Brown Argus*, Northern Brown Argus*

Cow wheat	Heath Fritillary*
Cowslip	Duke of Burgundy*
Devil's bit scabious	Marsh Fritillary*
Docks	Small Copper
Dogwood	Holly Blue
Elm (wych & common)	White-letter Hairstreak, Large Tortoiseshell*, Comma
Garlic mustard	Green-veined White, Orange Tip
Gorse	Silver-studded Blue*, Green Hairstreak*
Great water dock	Large Copper*
Heathers	Silver-studded Blue*
Hedge mustard	Orange Tip
Holly	Holly Blue
Honesty	Orange Tip
Honeysuckle	White Admiral*
Hop	Comma, Red Admiral
Horseshoe vetch	Chalkhill Blue*, Adonis Blue*
Ivy	Holly Blue
Kidney vetch	Small Blue*, Chalkhill Blue*
Ladies smock (cuckoo flower)	Orange Tip*
Lucerne	Clouded Yellow*
Milk parsley	Swallowtail*
Nasturtium	Large White, Small White
Oak	Purple Hairstreak
Plantains (ribwort & sea)	Glanville Fritillary*
Primrose	Duke of Burgundy*
Purging buckthorn	Brimstone
Red clover	Common Blue, Clouded Yellow*
Rest harrow	Common Blue
Rockrose	Brown Argus*, Northern Brown Argus*
Sallow (broad leaved)	Purple Emperor*, Comma
Sorrell (common & sheep)	Small Copper
Stinging nettle	Red Admiral, Small Tortoiseshell, Peacock, Comma, Painted Lady
Strawberry (barren & wild)	Grizzled Skipper*
Sweet rocket	Orange Tip
Thistles	Painted Lady

Plants for Egg Laying and Caterpillar Feeding—continued

Thyme	Large Blue*, Silver-studded Blue*
Tormentil	Common Blue
Vetch species	Wood White*
Violet (Dog & sweet)	Dark Green Fritillary*, Pearl-bordered Fritillary*, Small Pearl-bordered Fritillary*, High Brown Fritillary*, Silver-washed Fritillary*
Willow	Large Tortoiseshell*

Grasses used for eggs and by caterpillars

Grasses:

Annual meadow	Tufted hair	Early hair	Chalk false brome	Sheep's fescue
Couch	Yorkshire fog	Purple moor	Cock's foot	Wood false brome
Meadow fescue	Cat's tail	White beak sedge	Mat	Woodland meadow

. . . . used by these butterflies:

Essex Skipper	Small -mountain Ringlet*	Wall Brown	**Rye grass is not** used by any species of British butterfly.
Silver-spotted Skipper*	Gatekeeper	Marbled White	
Lulworth Skipper*	Scotch Argus*	Ringlet	
Meadow Brown	Speckled Wood	Grayling*	
	Small Heath	Large Heath*	
Chequered Skipper*	Large Skipper	Small Skipper	

*Unlikely to visit gardens because of specialised habitat requirement or rarity.

Appendix C

SELECTED BOOK LIST

Butterfly Books

Brooks & Knight (Field Guide)	A Complete Guide to British Butterflies - Jonathon Cape 1982
Gooden R.	Butterflies of Britain & Europe - Collins
	British Butterflies - a Field Guide - David and Charles 1978
Heath, Pollard & Thomas	Atlas of Butterflies in Britain and Ireland - Viking 1984
Thomas J.A. (First Nature)	RSNC Guide to Butterflies of the British Isles - Hamlyn 1986
	Butterflies and Moths - Usborne 1980 (for children)

Selected Book List — continued

Butterfly Gardening and Breeding

Oates M.	Garden Plants for Butterflies - Masterton 1985
Friedrich E.	Breeding Butterflies and Moths - Harley Books
Rothschild & Farrell	The Butterfly Gardener - Michael Joseph 1983
Cribb P.W.	Breeding the British Butterflies - Amateur Entomological Society 1983 (355 Hounslow Rd. Hanworth, Feltham, Middlesex, TW13 5JH)
Butterfly News	bi-monthly (Editor: Simon Regan, Lodmoor Country Park, Greenhill, Weymouth, Dorset)
Baines Chris	How to make a Wildlife Garden - Elmtree Books 1985

Wildflower Books

W. Keble Martin	The Concise British Flora in Colour - Ebury Press & Michael Joseph 1965
Phillips R.	Wild Flowers of Britain - Pan 1977
Stevens J.	Nat. Trust Book of Wild Flower Gardening - Dorling Kindersley 1987

Appendix D

USEFUL RESOURCES

From **Butterfly Park, Lincs.:**
Breeding cages (ideal for classroom or home)
British livestock (subject to availability)
Landscaping design and gardening services

From **School Garden Company:**
Full range of butterfly & gardening books in print
Wild flower, grasses and mixed seeds
Butterfly posters, slides and children's books
... please ring or write for a catalogue.

Suppliers of **Sapporo Autumn Gold** disease resistant elms:
Crowders Nurseries, Lincoln Road, Horncastle, Lincs.
Elms Across Europe - Pitney Bowes plc., The Pinnacles, Harlow, Essex, CM19 5BD.

Appendix E

CONTACT ADDRESSES

Butterfly Park, Long Sutton, Lincolnshire. PE12 9LE. Tel. 0406 - 363833

School Garden Company, P.O. Box 49, Spalding, Lincs. PE11 1NZ. Tel. 0775 - 69518

British Butterfly Conservation Society, Tudor House, Quorn, Nr. Loughborough, Leics. LE12 8AD.
The only society solely concerned with the conservation of British butterflies - anyone interested in butterflies should certainly join. The Society produces an excellent bulletin - BBCS News - and has many active Regional Branches.

School Natural Science Society - membership sec.: A. Nicholls, 9 Killington Drive, Kendal, Cumbria. LA9 7NY.

Nature Conservancy Council, Northminster House, Peterborough. PE1 1UA.

Royal Society for Nature Conservation, The Green, Nettleham, Lincoln. LN2 2NR.

CLEAPSE School Science Service, Brunel University, Uxbridge. UB8 3PH.

Council for Environmental Education, School of Education, University of Reading, London Road, Reading. RG1 5AQ.

Conservation Trust, George Palmer Site, Northumberland Avenue, Reading. RG2 7PW.